THE EDGE OF THE KNIFE

By

H. BEAM PIPER

dition et mise en page : Culturea (Hérault, 34)
etrouvez notre catalogue sur http://culturea.fr/
ontact : infos@culturea.fr
nprimé en Allemagne par Books on Demand Gmbh
esign typographique et layout : Derek Murphy et Reedsy
SBN : 9791041985449
ate de parution : février 2024
e livre :
– a été composé avec une police open source libre de droits dessinée sur Open Fondry https://open-foundry.com/
– est édité en libre access sous licence CC0 (Creative Commons Zero)
afin de conserver l'oeuvre au plus près des caractéristiques du domaine public.
– offre la possibilité à son lecteur de partager, dupliquer ou distribuer son contenu, sans restriction ni réserve.
– permet de replanter un arbre. SHS Éditions soutient Plantons Pour l'Avenir, fonds de dotation pour le reboisement des forêts françaises
400 projets de reboisement de forêts soutenus depuis 2014 à travers la France https://www.plantonspourlavenir.fr/

This story was rejected by two top-flight science-fiction editors for the same reason: "Too hot to handle." "Too dangerous for our book." We'd like to know whether or not the readers of Amazing Stories *agree.*

Chalmers stopped talking abruptly, warned by the sudden attentiveness of the class in front of him. They were all staring; even Guellick, in the fourth row, was almost half awake. Then one of them, taking his silence as an invitation to questions found his voice.

"You say Khalid ib'n Hussein's been assassinated?" he asked incredulously. "When did that happen?"

There was no past—no future—only a great chaotic NOW.

"In 1973, at Basra." There was a touch of impatience in his voice; surely they ought to know that much. "He was shot, while leaving the Parliament Building, by an Egyptian Arab named Mohammed Noureed, with an old U. S. Army M3 submachine-gun. Noureed killed two of Khalid's guards and wounded another before he was overpowered. He was lynched on the spot by the crowd; stoned to death. Ostensibly, he and his accomplices were religious fanatics; however, there can be no doubt whatever that the murder was inspired, at least indirectly, by the Eastern Axis."

The class stirred like a grain-field in the wind. Some looked at him in blank amazement; some were hastily averting faces red with poorly suppressed laughter. For a moment he was puzzled, and then realization hit him like a blow in the stomach-pit. He'd forgotten, again.

"I didn't see anything in the papers about it," one boy was saying.

"The newscast, last evening, said Khalid was in Ankara, talking to the President of Turkey," another offered.

"Professor Chalmers, would you tell us just what effect Khalid's death had upon the Islamic Caliphate and the Middle Eastern situation in general?" a third voice asked with exaggerated solemnity. That was Kendrick, the class humorist; the question was pure baiting.

"Well, Mr. Kendrick, I'm afraid it's a little too early to assess the full results of a thing like that, if they can ever be fully assessed. For instance, who, in 1911, could have predicted all the consequences of the pistol-shot at Sarajevo? Who, even today, can guess what the history of the world would have been had Zangarra not missed Franklin Roosevelt in 1932? There's always that if."

He went on talking safe generalities as he glanced covertly at his watch. Only five minutes to the end of the period; thank heaven he hadn't made that slip at the beginning of the class. "For instance, tomorrow, when we take up the events in India from the First World War to the end of British rule, we will be largely concerned with another victim of the assassin's bullet, Mohandas K. Gandhi. You may ask yourselves, then, by how much that bullet altered the history of the Indian sub-continent. A word of warning, however: The events we will be discussing will be either contemporary with or prior to what was discussed today. I hope that you're all keeping your notes properly dated. It's always easy to become confused in matters of chronology."

He wished, too late, that he hadn't said that. It pointed up the very thing he was trying to play down, and raised a general laugh.

As soon as the room was empty, he hastened to his desk, snatched pencil and notepad. This had been a bad one, the worst yet; he hadn't heard the end of it by any means. He couldn't waste thought on that now, though. This was all new and important; it had welled up suddenly and without warning into his conscious mind, and he must get it down in notes before the "memory"—even mentally, he always put that word into quotes—was lost. He was still scribbling furiously when the instructor who would use the room for the next period entered, followed by a few of his students. Chalmers finished, crammed the notes into his pocket, and went out into the hall.

Most of his own Modern History IV class had left the building and were on their way across the campus for science classes. A few, however, were joining groups for other classes here in Prescott Hall, and in every group, they were the center of interest. Sometimes, when they saw him, they would fall silent until he had passed; sometimes they didn't, and he caught snatches of conversation.

"Oh, brother! Did Chalmers really blow his jets this time!" one voice was saying.

"Bet he won't be around next year."

Another quartet, with their heads together, were talking more seriously.

"Well, I'm not majoring in History, myself, but I think it's an outrage that some people's diplomas are going to depend on grades given by a lunatic!"

"Mine will, and I'm not going to stand for it. My old man's president of the Alumni Association, and...."

That was something he had not thought of, before. It gave him an ugly start. He was still thinking about it as he turned into the side hall to the History Department offices and entered the cubicle he shared with a colleague. The colleague, old Pottgeiter, Medieval History, was emerging in a rush; short, rotund, gray-bearded, his arms full of books and papers, oblivious, as usual, to anything that had happened since the Battle of Bosworth or the Fall of Constantinople. Chalmers stepped quickly out of his way and entered behind him. Marjorie Fenner, the secretary they also shared, was tidying up the old man's desk.

"Good morning, Doctor Chalmers." She looked at him keenly for a moment. "They give you a bad time again in Modern Four?"

Good Lord, did he show it that plainly? In any case, it was no use trying to kid Marjorie. She'd hear the whole story before the end of the day.

"Gave myself a bad time."

Marjorie, still fussing with Pottgeiter's desk, was about to say something in reply. Instead, she exclaimed in exasperation.

"Ohhh! That man! He's forgotten his notes again!" She gathered some papers from Pottgeiter's desk, rushing across the room and out the door with them.

For a while, he sat motionless, the books and notes for General European History II untouched in front of him. This was going to raise hell. It hadn't been the first slip he'd made, either; that thought kept recurring to him. There had been the time when he had alluded to the colonies on Mars and Venus. There had been the time he'd mentioned the secession of Canada from the British Commonwealth, and the time he'd called the U. N. the Terran Federation. And the time he'd tried to get a copy of Franchard's *Rise and Decline of the System States*, which wouldn't be published until the Twenty-eighth Century, out of the college library. None of those had drawn much comment, beyond a few student jokes about the history professor who lived in the future instead of the past. Now, however, they'd all be remembered, raked up, exaggerated, and added to what had happened this morning.

He sighed and sat down at Marjorie's typewriter and began transcribing his notes. Assassination of Khalid ib'n Hussein, the pro-Western leader of the newly formed Islamic Caliphate; period of anarchy in the Middle East; interfactional power-struggles; Turkish intervention. He wondered how long that would last; Khalid's son, Tallal ib'n Khalid, was at school in England when his father was—would be—killed. He would return, and eventually take his father's place, in time to bring the Caliphate into the Terran Federation when the general war came. There were some notes on that already; the war would result from an attempt by the Indian Communists to seize East Pakistan. The trouble was that he so seldom "remembered" an exact date. His "memory" of the year of Khalid's assassination was an exception.

Nineteen seventy-three—why, that was this year. He looked at the calendar. October 16, 1973. At very most, the Arab statesman had two and a half months to live. Would there be any possible way in which he could give a credible warning? He doubted it. Even if there were, he questioned whether he should—for that matter, whether he *could*—interfere....

He always lunched at the Faculty Club; today was no time to call attention to himself by breaking an established routine. As he entered, trying to avoid either a furtive slink or a chip-on-shoulder swagger, the crowd in the lobby stopped talking abruptly, then began again on an obviously changed subject. The word had gotten around, apparently. Handley, the head of the Latin Department, greeted him with a distantly polite nod. Pompous old owl; regarded himself, for some reason, as a sort of unofficial Dean of the Faculty. Probably didn't want to be seen fraternizing with controversial characters. One of the younger men, with a thin face and a mop of unruly hair, advanced to meet him as he came in, as cordial as Handley was remote.

"Oh, hello, Ed!" he greeted, clapping a hand on Chalmers' shoulder. "I was hoping I'd run into you. Can you have dinner with us this evening?" He was sincere.

"Well, thanks, Leonard. I'd like to, but I have a lot of work. Could you give me a rain-check?"

"Oh, surely. My wife was wishing you'd come around, but I know how it is. Some other evening?"

"Yes, indeed." He guided Fitch toward the dining-room door and nodded toward a table. "This doesn't look too crowded; let's sit here."

After lunch, he stopped in at his office. Marjorie Fenner was there, taking dictation from Pottgeiter; she nodded to him as he entered, but she had no summons to the president's office.

The summons was waiting for him, the next morning, when he entered the office after Modern History IV, a few minutes past ten.

"Doctor Whitburn just phoned," Marjorie said. "He'd like to see you, as soon as you have a vacant period."

"Which means right away. I shan't keep him waiting."

She started to say something, swallowed it, and then asked if he needed anything typed up for General European II.

"No, I have everything ready." He pocketed the pipe he had filled on entering, and went out.

The president of Blanley College sat hunched forward at his desk; he had rounded shoulders and round, pudgy fists and a round, bald head. He seemed to be expecting his visitor to stand at attention in front of him. Chalmers got the pipe out of his pocket, sat down in the desk-side chair, and snapped his lighter.

"Good morning, Doctor Whitburn," he said very pleasantly.

Whitburn's scowl deepened. "I hope I don't have to tell you why I wanted to see you," he began.

"I have an idea." Chalmers puffed until the pipe was drawing satisfactorily. "It might help you get started if you did, though."

"I don't suppose, at that, that you realize the full effect of your performance, yesterday morning, in Modern History Four," Whitburn replied. "I don't suppose you know, for instance, that I had to intervene at the last moment and suppress an editorial in the *Black and Green*, derisively critical of you and your teaching methods, and, by implication, of the administration of this college. You didn't hear about that, did you? No, living as you do in the future, you wouldn't."

"If the students who edit the *Black and Green* are dissatisfied with anything here, I'd imagine they ought to say so," Chalmers commented. "Isn't that what they teach in the journalism classes, that the purpose of journalism is to speak for the dissatisfied? Why make exception?"

"I should think you'd be grateful to me for trying to keep your behavior from being made a subject of public ridicule among your students. Why, this editorial which I suppressed actually went so far as to question your sanity!"

"I should suppose it might have sounded a good deal like that, to them. Of course, I have been preoccupied, lately, with an imaginative projection of present trends into the future. I'll quite freely admit that I should have kept my extracurricular work separate from my class and lecture work, but...."

"That's no excuse, even if I were sure it were true! What you did, while engaged in the serious teaching of history, was to indulge in a farrago of nonsense, obvious as such to any child, and damage not only your own standing with your class but the standing of Blanley College as well. Doctor Chalmers, if this were the first incident of the kind it would be bad enough, but it isn't. You've done things like this before, and I've warned you before. I assumed, then, that you were merely showing the effects of overwork, and I offered you a vacation, which you refused to take. Well, this is the limit. I'm compelled to request your immediate resignation."

Chalmers laughed. "A moment ago, you accused me of living in the future. It seems you're living in the past. Evidently you haven't heard about the Higher Education Faculty Tenure Act of 1963, or such things as tenure-contracts. Well, for your information, I have one; you signed it yourself, in case you've forgotten. If you want my resignation, you'll have to show cause, in a court of law, why my contract should be voided, and I don't think a slip of the tongue is a reason for voiding a contract that any court would accept."

Whitburn's face reddened. "You don't, don't you? Well, maybe it isn't, but insanity is. It's a very good reason for voiding a contract voidable on grounds of unfitness or incapacity to teach."

He had been expecting, and mentally shrinking from, just that. Now that it was out, however, he felt relieved. He gave another short laugh.

"You're willing to go into open court, covered by reporters from papers you can't control as you do this student sheet here, and testify that for the past twelve years you've had an insane professor on your faculty?"

"You're.... You're trying to blackmail me?" Whitburn demanded, half rising.

"It isn't blackmail to tell a man that a bomb he's going to throw will blow up in his hand." Chalmers glanced quickly at his watch. "Now, Doctor Whitburn, if you have nothing further to discuss, I have a class in a few minutes. If you'll excuse me...."

He rose. For a moment, he stood facing Whitburn; when the college president said nothing, he inclined his head politely and turned, going out.

Whitburn's secretary gave the impression of having seated herself hastily at her desk the second before he opened the door. She watched him, round-eyed, as he went out into the hall.

He reached his own office ten minutes before time for the next class. Marjorie was typing something for Pottgeiter; he merely nodded to her, and picked up the phone. The call would have to go through the school exchange, and he had a suspicion that Whitburn kept a check on outside calls. That might not hurt any, he thought, dialing a number.

"Attorney Weill's office," the girl who answered said.

"Edward Chalmers. Is Mr. Weill in?"

She'd find out. He was; he answered in a few seconds.

"Hello, Stanly; Ed Chalmers. I think I'm going to need a little help. I'm having some trouble with President Whitburn, here at the college. A matter involving the validity of my tenure-contract. I don't want to go into it over this line. Have you anything on for lunch?"

"No, I haven't. When and where?" the lawyer asked.

He thought for a moment. Nowhere too close the campus, but not too far away.

"How about the Continental; Fontainbleu Room? Say twelve-fifteen."

"That'll be all right. Be seeing you."

Marjorie looked at him curiously as he gathered up the things he needed for the next class.

Stanly Weill had a thin dark-eyed face. He was frowning as he set down his coffee-cup.

"Ed, you ought to know better than to try to kid your lawyer," he said. "You say Whitburn's trying to force you to resign. With your contract, he can't do that, not without good and sufficient cause, and under the Faculty Tenure Law, that means something just an inch short of murder in the first degree. Now, what's Whitburn got on you?"

Beat around the bush and try to build a background, or come out with it at once and fill in the details afterward? He debated mentally for a moment, then decided upon the latter course.

"Well, it happens that I have the ability to prehend future events. I can, by concentrating, bring into my mind the history of the world, at least in general outline, for the next five thousand years. Whitburn thinks I'm crazy, mainly because I get confused at times and forget that something I know about hasn't happened yet."

Weill snatched the cigarette from his mouth to keep from swallowing it. As it was, he choked on a mouthful of smoke and coughed violently, then sat back in the booth-seat, staring speechlessly.

"It started a little over three years ago," Chalmers continued. "Just after New Year's, 1970. I was getting up a series of seminars for some of my postgraduate students on extrapolation of present social and political trends to the middle of the next century, and I began to find that I was getting some very fixed and definite ideas of what the world of 2050 to 2070 would be like. Completely unified world, abolition of all national states under a single world sovereignty, colonies on Mars and Venus, that sort of thing. Some of these ideas didn't seem quite logical; a number of them were complete reversals of present trends, and a lot seemed to depend on arbitrary and unpredictable factors. Mind, this was before the first rocket landed on the Moon, when the whole moon-rocket and lunar-base project was a triple-top secret. But I knew, in the spring of 1970, that the first unmanned rocket would be called the *Kilroy*, and that it would be launched some time in 1971. You remember, when the news was released, it was stated that the rocket hadn't been christened until the day before it was launched, when somebody remembered that old 'Kilroy-was-here' thing from the Second World War. Well, I knew about it over a year in advance."

Weill had been listening in silence. He had a naturally skeptical face; his present expression mightn't really mean that he didn't believe what he was hearing.

"How'd you get all this stuff? In dreams?"

Chalmers shook his head. "It just came to me. I'd be sitting reading, or eating dinner, or talking to one of my classes, and the first thing I'd know, something out of the future would come bubbling up in me. It just kept pushing up into my conscious mind. I wouldn't have an idea of something one minute, and the next it would just be part of my general historical knowledge; I'd know it as positively as I know that Columbus discovered America in. 1492. The only difference is that I can usually remember where I've read something in past history, but my future history I know without knowing how I know it."

"Ah, that's the question!" Weill pounced. "You don't know how you know it. Look, Ed, we've both studied psychology, elementary psychology at least. Anybody who has to work with people, these days, has to know some psychology. What makes you sure that these prophetic impressions of yours aren't manufactured in your own subconscious mind?"

"That's what I thought, at first. I thought my subconscious was just building up this stuff to fill the gaps in what I'd produced from logical extrapolation. I've always been a stickler for detail," he added, parenthetically. "It would be natural for me to supply details for the future. But, as I said, a lot of this stuff is based on unpredictable and arbitrary factors that can't be inferred from anything in the present. That left me with the alternatives of delusion or precognition, and if I ever came near going crazy, it was before the *Kilroy* landed and the news was released. After that, I knew which it was."

"And yet, you can't explain how you can have real knowledge of a thing before it happens. Before it exists," Weill said.

"I really don't need to. I'm satisfied with knowing that I know. But if you want me to furnish a theory, let's say that all these things really do exist, in the past or in the future, and that the present is just a moving knife-edge that separates the two. You can't even indicate the present. By the time you make up your mind to say, 'Now!' and transmit the impulse to your vocal organs, and utter the word, the original present moment is part of the past. The knife-edge has gone over it. Most people think they know only the present; what they know is the past, which they have already experienced, or read about. The difference with me is that I can see what's on both sides of the knife-edge."

Weill put another cigarette in his mouth and bent his head to the flame of his lighter. For a moment, he sat motionless, his thin face rigid.

"What do you want me to do?" he asked. "I'm a lawyer, not a psychiatrist."

"I want a lawyer. This is a legal matter. Whitburn's talking about voiding my tenure contract. You helped draw it; I have a right to expect you to help defend it."

"Ed, have you been talking about this to anybody else?" Weill asked.

"You're the first person I've mentioned it to. It's not the sort of thing you'd bring up casually, in a conversation."

"Then how'd Whitburn get hold of it?"

"He didn't, not the way I've given it to you. But I made a couple of slips, now and then. I made a bad one yesterday morning."

He told Weill about it, and about his session with the president of the college that morning. The lawyer nodded.

"That was a bad one, but you handled Whitburn the right way," Weill said. "What he's most afraid of is publicity, getting the college mixed up in anything controversial, and above all, the reactions of the trustees and people like that. If Dacre or anybody else makes any trouble, he'll do his best to cover for you. Not willingly, of course, but because he'll know that that's the only way he can cover for himself. I don't think you'll have any more trouble with him. If you can keep your own nose clean, that is. Can you do that?"

"I believe so. Yesterday I got careless. I'll not do that again."

"You'd better not." Weill hesitated for a moment. "I said I was a lawyer, not a psychiatrist. I'm going to give you some psychiatrist's advice, though. Forget this whole thing. You say you can bring these impressions into your conscious mind by concentrating?" He waited briefly; Chalmers nodded, and he continued: "Well, stop it. Stop trying to harbor this stuff. It's dangerous, Ed. Stop playing around with it."

"You think I'm crazy, too?"

Weill shook his head impatiently. "I didn't say that. But I'll say, now, that you're losing your grip on reality. You are constructing a system of fantasies, and the first thing you know, they will become your reality, and the world around you will be unreal and illusory. And that's a state of mental incompetence that I can recognize, as a lawyer."

"How about the *Kilroy*?"

Weill looked at him intently. "Ed, are you sure you did have that experience?" he asked. "I'm not trying to imply that you're consciously lying to me about that. I am suggesting that you manufactured a memory of that incident in your subconscious mind, and are deluding yourself into thinking that you knew about it in advance. False memory is a fairly common thing, in cases like this. Even the little psychology I know, I've heard about that. There's been talk about rockets to the Moon for years. You included something about that in your future-history fantasy, and then, after the event, you convinced yourself that you'd known all about it, including the impromptu christening of the rocket, all along."

A hot retort rose to his lips; he swallowed it hastily. Instead, he nodded amicably.

"That's a point worth thinking of. But right now, what I want to know is, will you represent me in case Whitburn does take this to court and does try to void my contract?"

"Oh, yes; as you said, I have an obligation to defend the contracts I draw up. But you'll have to avoid giving him any further reason for trying to void it. Don't make any more of these slips. Watch what you say, in class or out of it. And above all, don't talk about this to anybody. Don't tell anybody that you can foresee the future, or even talk about future probabilities. Your business is with the past; stick to it."

The afternoon passed quietly enough. Word of his defiance of Whitburn had gotten around among the faculty—Whitburn might have his secretary scared witless in his office, but not gossipless outside it—though it hadn't seemed to have leaked down to the students yet. Handley, the Latin professor, managed to waylay him in a hallway, a hallway Handley didn't normally use.

"The tenure-contract system under which we hold our positions here is one of our most valuable safeguards," he said, after exchanging greetings. "It was only won after a struggle, in a time of public animosity toward all intellectuals, and even now, our professional position would be most insecure without it."

"Yes. I found that out today, if I hadn't known it when I took part in the struggle you speak of."

"It should not be jeopardized," Handley declared.

"You think I'm jeopardizing it?"

Handley frowned. He didn't like being pushed out of the safety of generalization into specific cases.

"Well, now that you make that point, yes. I do. If Doctor Whitburn tries to make an issue of ... of what happened yesterday ... and if the court decides against you, you can see the position all of us will be in."

"What do you think I should have done? Given him my resignation when he demanded it? We have our tenure-contracts, and the system was instituted to prevent just the sort of arbitrary action Whitburn tried to take with me today. If he wants to go to court, he'll find that out."

"And if he wins, he'll establish a precedent that will threaten the security of every college and university faculty member in the state. In any state where there's a tenure law."

Leonard Fitch, the psychologist, took an opposite attitude. As Chalmers was leaving the college at the end of the afternoon, Fitch cut across the campus to intercept him.

"I heard about the way you stood up to Whitburn this morning, Ed," he said. "Glad you did it. I only wish I'd done something like that three years ago.... Think he's going to give you any real trouble?"

"I doubt it."

"Well, I'm on your side if he does. I won't be the only one, either."

"Well, thank you, Leonard. It always helps to know that. I don't think there'll be any more trouble, though."

He dined alone at his apartment, and sat over his coffee, outlining his work for the next day. When both were finished, he dallied indecisively, Weill's words echoing through his mind and raising doubts. It was possible that he had been manufacturing the whole thing in his subconscious mind. That was, at least, a more plausible theory than any he had constructed to explain an ability to produce real knowledge of the future. Of course, there was that business about the *Kilroy.* That had been too close on too many points to be dismissed as coincidence. Then, again, Weill's words came back to disquiet him. Had he really gotten that before the event, as he believed, or had he only imagined, later, that he had?

There was one way to settle that. He rose quickly and went to the filing-cabinet where he kept his future-history notes and began pulling out envelopes. There was nothing about the *Kilroy* in the Twentieth Century file, where it should be, although he examined each sheet of notes carefully. The possibility that his notes on that might have been filed out of place by mistake occurred to him; he looked in every other envelope. The notes, as far as they went, were all filed in order, and each one bore, beside the future date of occurrence, the date on which the knowledge—or must he call it delusion?—had come to him. But there was no note on the landing of the first unmanned rocket on Luna.

He put the notes away and went back to his desk, rummaging through the drawers, and finding nothing. He searched everywhere in the apartment where a sheet of paper could have been mislaid, taking all his books, one by one, from the shelves and leafing through them, even books he knew he had not touched for more than three years. In the end, he sat down again at his desk, defeated. The note on the *Kilroy* simply did not exist.

Of course, that didn't settle it, as finding the note would have. He remembered—or believed he remembered—having gotten that item of knowledge—or delusion—in

1970, shortly before the end of the school term. It hadn't been until after the fall opening of school that he had begun making notes. He could have had the knowledge of the robot rocket in his mind then, and neglected putting it on paper.

He undressed, put on his pajamas, poured himself a drink, and went to bed. Three hours later, still awake, he got up, and poured himself another, bigger, drink. Somehow, eventually, he fell asleep.

The next morning, he searched his desk and book-case in the office at school. He had never kept a diary; now he was wishing that he had. That might have contained something that would be evidence, one way or the other. All day, he vacillated between conviction of the reality of his future knowledge and resolution to have no more to do with it. Once he decided to destroy all the notes he had made, and thought of making a special study of some facet of history, and writing another book, to occupy his mind.

After lunch, he found that more data on the period immediately before the Thirty Days' War was coming into his consciousness. He resolutely suppressed it, knowing as he did that it might never come to him again. That evening, too, he cooked dinner for himself at his apartment, and laid out his class-work for the next day. He'd better not stay in, that evening; too much temptation to settle himself by the living-room fire with his pipe and his notepad and indulge in the vice he had determined to renounce. After a little debate, he decided upon a movie; he put on again the suit he had taken off on coming home, and went out.

The picture, a random choice among the three shows in the neighborhood, was about Seventeenth Century buccaneers; exciting action and a sound-track loud with shots and cutlass-clashing. He let himself be drawn into it completely, and, until it was finished, he was able to forget both the college and the history of the future. But, as he walked home, he was struck by the parallel between the buccaneers of the West Indies and the space-pirates in the days of the dissolution of the First Galactic Empire, in the Tenth Century of the Interstellar Era. He hadn't been too clear on that period, and he found new data rising in his mind; he hurried his steps, almost running upstairs to his room. It was long after midnight before he had finished the notes he had begun on his return home.

Well, that had been a mistake, but he wouldn't make it again. He determined again to destroy his notes, and began casting about for a subject which would occupy his mind to the exclusion of the future. Not the Spanish Conquistadores; that was too much like the early period of interstellar expansion. He thought for a time of the Sepoy Mutiny, and then rejected it—he could "remember" something much like that on one of the planets of the Beta Hydrae system, in the Fourth Century of the Atomic Era. There were so few things, in the history of the past, which did not have their counter-parts in the future. That evening, too, he stayed at home, preparing for his various classes for the rest of the week and making copious notes on what he would talk about to each. He needed more whiskey to get to sleep that night.

Whitburn gave him no more trouble, and if any of the trustees or influential alumni made any protest about what had happened in Modern History IV, he heard nothing about it. He managed to conduct his classes without further incidents, and spent his evenings trying, not always successfully, to avoid drifting into "memories" of the future....

He came into his office that morning tired and unrefreshed by the few hours' sleep he had gotten the night before, edgy from the strain, of trying to adjust his mind to the world of Blanley College in mid-April of 1973. Pottgeiter hadn't arrived yet, but Marjorie Fenner was waiting for him; a newspaper in her hand, almost bursting with excitement.

"Here; have you seen it, Doctor Chalmers?" she asked as he entered.

He shook his head. He ought to read the papers more, to keep track of the advancing knife-edge that divided what he might talk about from what he wasn't supposed to know, but each morning he seemed to have less and less time to get ready for work.

"Well, look! Look at that!"

She thrust the paper into his hands, still folded, the big, black headline where he could see it.

KHALID IB'N HUSSEIN ASSASSINATED

He glanced over the leading paragraphs. Leader of Islamic Caliphate shot to death in Basra ... leaving Parliament Building for his palace outside the city ... fanatic, identified as an Egyptian named Mohammed Noureed ... old American submachine-gun ... two guards killed and a third seriously wounded ... seized by infuriated mob and stoned to death on the spot....

For a moment, he felt guilt, until he realized that nothing he could have done could have altered the event. The death of Khalid ib'n Hussein, and all the millions of other deaths that would follow it, were fixed in the matrix of the space-time continuum. Including, maybe, the death of an obscure professor of Modern History named Edward Chalmers.

"At least, this'll be the end of that silly flap about what happened a month ago in Modern Four. This is modern history, now; I can talk about it without a lot of fools yelling their heads off."

She was staring at him wide-eyed. No doubt horrified at his cold-blooded attitude toward what was really a shocking and senseless crime.

"Yes, of course; the man's dead. So's Julius Caesar, but we've gotten over being shocked at his murder."

He would have to talk about it in Modern History IV, he supposed; explain why Khalid's death was necessary to the policies of the Eastern Axis, and what the consequences would be. How it would hasten the complete dissolution of the old U. N., already weakened by the crisis over the Eastern demands for the demilitarization and internationalization of the United States Lunar Base, and necessitate the formation of the Terran Federation, and how it would lead, eventually, to the Thirty Days' War. No, he couldn't talk about that; that was on the wrong side of the knife-edge. Have to be careful about the knife-edge; too easy to cut himself on it.

Nobody in Modern History IV was seated when he entered the room; they were all crowded between the door and his desk. He stood blinking, wondering why they were giving him an ovation, and why Kendrick and Dacre were so abjectly apologetic. Great heavens, did it take the murder of the greatest Moslem since Saladin to convince people that he wasn't crazy?

Before the period was over, Whitburn's secretary entered with a note in the college president's hand and over his signature; requesting Chalmers to come to his office immediately and without delay. Just like that; expected him to walk right out of his class. He was protesting as he entered the president's office. Whitburn cut him off short.

"Doctor Chalmers,"—Whitburn had risen behind his desk as the door opened—"I certainly hope that you can realize that there was nothing but the most purely coincidental connection between the event featured in this morning's newspapers and your performance, a month ago, in Modern History Four," he began.

"I realize nothing of the sort. The death of Khalid ib'n Hussein is a fact of history, unalterably set in its proper place in time-sequence. It was a fact of history a month ago no less than today."

"So that's going to be your attitude; that your wild utterances of a month ago have now been vindicated as fulfilled prophesies? And I suppose you intend to exploit this—this coincidence—to the utmost. The involvement of Blanley College in a mess of sensational publicity means nothing to you, I presume."

"I haven't any idea what you're talking about."

"You mean to tell me that you didn't give this story to the local newspaper, the *Valley Times*?" Whitburn demanded.

"I did not. I haven't mentioned the subject to anybody connected with the *Times*, or anybody else, for that matter. Except my attorney, a month ago, when you were threatening to repudiate the contract you signed with me."

"I suppose I'm expected to take your word for that?"

"Yes, you are. Unless you care to call me a liar in so many words." He moved a step closer. Lloyd Whitburn outweighed him by fifty pounds, but most of the difference was fat. Whitburn must have realized that, too.

"No, no; if you say you haven't talked about it to the *Valley Times*, that's enough," he said hastily. "But somebody did. A reporter was here not twenty minutes ago; he refused to say who had given him the story, but he wanted to question me about it."

"What did you tell him?"

"I refused to make any statement whatever. I also called Colonel Tighlman, the owner of the paper, and asked him, very reasonably, to suppress the story. I thought that my own position and the importance of Blanley College to this town entitled me to that much consideration." Whitburn's face became almost purple. "He ... he laughed at me!"

"Newspaper people don't like to be told to kill stories. Not even by college presidents. That's only made things worse. Personally, I don't relish the prospect of having this publicized, any more than you do. I can assure you that I shall be most guarded if any of the *Times* reporters talk to me about it, and if I have time to get back to my class before the end of the period, I shall ask them, as a personal favor, not to discuss the matter outside."

Whitburn didn't take the hint. Instead, he paced back and forth, storming about the reporter, the newspaper owner, whoever had given the story to the paper, and finally Chalmers himself. He was livid with rage.

"You certainly can't imagine that when you made those remarks in class you actually possessed any knowledge of a thing that was still a month in the future," he spluttered. "Why, it's ridiculous! Utterly preposterous!"

"Unusual, I'll admit. But the fact remains that I did. I should, of course, have been more careful, and not confused future with past events. The students didn't understand...."

Whitburn half-turned, stopping short.

"My God, man! You *are* crazy!" he cried, horrified.

The period-bell was ringing as he left Whitburn's office; that meant that the twenty-three students were scattering over the campus, talking like mad. He shrugged. Keeping them quiet about a thing like this wouldn't have been possible in any case. When he entered his office, Stanly Weill was waiting for him. The lawyer drew him out into the hallway quickly.

"For God's sake, have you been talking to the papers?" he demanded. "After what I told you...."

"No, but somebody has." He told about the call to Whitburn's office, and the latter's behavior. Weill cursed the college president bitterly.

"Any time you want to get a story in the *Valley Times*, just order Frank Tighlman not to print it. Well, if you haven't talked, don't."

"Suppose somebody asks me?"

"A reporter, no comment. Anybody else, none of his damn business. And above all, don't let anybody finagle you into making any claims about knowing the future. I thought we had this under control; now that it's out in the open, what that fool Whitburn'll do is anybody's guess."

Leonard Fitch met him as he entered the Faculty Club, sizzling with excitement.

"Ed, this has done it!" he began, jubilantly. "This is one nobody can laugh off. It's direct proof of precognition, and because of the prominence of the event, everybody will hear about it. And it simply can't be dismissed as coincidence...."

"Whitburn's trying to do that."

"Whitburn's a fool if he is," another man said calmly. Turning, he saw that the speaker was Tom Smith, one of the math professors. "I figured the odds against that being chance. There are a lot of variables that might affect it one way or another, but ten to the fifteenth power is what I get for a sort of median figure."

"Did you give that story to the *Valley Times?*" he asked Fitch, suspicion rising and dragging anger up after it.

"Of course, I did," Fitch said. "I'll admit, I had to go behind your back and have some of my postgrads get statements from the boys in your history class, but you wouldn't talk about it yourself...."

Tom Smith was standing beside him. He was twenty years younger than Chalmers, he was an amateur boxer, and he had good reflexes. He caught Chalmers' arm as it was traveling back for an uppercut, and held it.

"Take it easy, Ed; you don't want to start a slugfest in here. This is the Faculty Club; remember?"

"I won't, Tom; it wouldn't prove anything if I did." He turned to Fitch. "I won't talk about sending your students to pump mine, but at least you could have told me before you gave that story out."

"I don't know what you're sore about," Fitch defended himself. "I believed in you when everybody else thought you were crazy, and if I hadn't collected signed and dated statements from your boys, there'd have been no substantiation. It happens that extrasensory perception means as much to me as history does to you. I've believed in it ever since I read about Rhine's work, when I was a kid. I worked in ESP for a long time. Then I had a chance to get a full professorship by coming here, and after I did, I found that I couldn't go on with it, because Whitburn's president here, and he's a stupid old bigot with an air-locked mind...."

"Yes." His anger died down as Fitch spoke. "I'm glad Tom stopped me from making an ass of myself. I can see your side of it." Maybe that was the curse of the professional intellectual, an ability to see everybody's side of everything. He thought for a moment. "What else did you do, beside hand this story to the *Valley Times?* I'd better hear all about it."

"I phoned the secretary of the American Institute of Psionics and Parapsychology, as soon as I saw this morning's paper. With the time-difference to the East Coast, I got him just as he reached his office. He advised me to give the thing the widest possible publicity; he thought that would advance the recognition and study of parapsychology. A case like this can't be ignored; it will demand serious study...."

"Well, you got your publicity, all right. I'm up to my neck in it."

There was an uproar outside. The doorman was saying, firmly:

"This is the Faculty Club, gentlemen; it's for members only. I don't care if you gentlemen are the press, you simply cannot come in here."

"We're all up to our necks in it," Smith said. "Leonard, I don't care what your motives were, you ought to have considered the effect on the rest of us first."

"This place will be a madhouse," Handley complained. "How we're going to get any of these students to keep their minds on their work...."

"I tell you, I don't know a confounded thing about it," Max Pottgeiter's voice rose petulantly at the door. "Are you trying to tell me that Professor Chalmers murdered some Arab? Ridiculous!"

He ate hastily and without enjoyment, and slipped through the kitchen and out the back door, cutting between two frat-houses and circling back to Prescott Hall. On the way, he paused momentarily and chuckled. The reporters, unable to storm the Faculty Club, had gone off in chase of other game and had cornered Lloyd Whitburn in front of Administration Center. They had a jeep with a sound-camera mounted on it, and were trying to get something for telecast. After gesticulating angrily, Whitburn broke away from them and dashed up the steps and into the building. A campus policeman stopped those who tried to follow.

His only afternoon class was American History III. He got through it somehow, though the class wasn't able to concentrate on the Reconstruction and the first election of Grover Cleveland. The halls were free of reporters, at least, and when it was over he hurried to the Library, going to the faculty reading-room in the rear, where he could smoke. There was nobody there but old Max Pottgeiter, smoking a cigar, his head bent over a book. The Medieval History professor looked up.

"Oh, hello, Chalmers. What the deuce is going on around here? Has everybody gone suddenly crazy?" he asked.

"Well, they seem to think I have," he said bitterly.

"They do? Stupid of them. What's all this about some Arab being shot? I didn't know there were any Arabs around here."

"Not here. At Basra." He told Pottgeiter what had happened.

"Well! I'm sorry to hear about that," the old man said. "I have a friend at Southern California, Bellingham, who knew Khalid very well. Was in the Middle East doing some research on the Byzantine Empire; Khalid was most helpful. Bellingham was quite impressed by him; said he was a wonderful man, and a fine scholar. Why would anybody want to kill a man like that?"

He explained in general terms. Pottgeiter nodded understandingly: assassination was a familiar feature of the medieval political landscape, too. Chalmers went on to elaborate. It was a relief to talk to somebody like Pottgeiter, who wasn't bothered by the present moment, but simply boycotted it. Eventually, the period-bell rang. Pottgeiter looked at his watch, as from conditioned reflex, and then rose, saying that he had a class and excusing himself. He would have carried his cigar with him if Chalmers hadn't taken it away from him.

After Pottgeiter had gone Chalmers opened a book—he didn't notice what it was—and sat staring unseeing at the pages. So the moving knife-edge had come down on the end of Khalid ib'n Hussein's life; what were the events in the next segment of time, and the segments to follow? There would be bloody fighting all over the Middle East—with consternation, he remembered that he had been talking about that to Pottgeiter. The Turkish army would move in and try to restore order. There would be more trouble in northern Iran, the Indian Communists would invade Eastern Pakistan, and then the general war, so long dreaded, would come. How far in the future that was he could not "remember," nor how the nuclear-weapons stalemate that had so far prevented it would be broken. He knew that today, and for years before, nobody had dared start an all-out atomic war. Wars, now, were marginal skirmishes, like the one in Indonesia, or the steady underground conflict of subversion and sabotage that had come to be called the Subwar. And with the United States already in possession of a powerful Lunar base.... He wished he could "remember" how events between the murder of Khalid and the Thirty Day's War had been spaced chronologically. Something of that had come to him, after the incident in Modern History IV, and he had driven it from his consciousness.

He didn't dare go home where the reporters would be sure to find him. He simply left the college, at the end of the school-day, and walked without conscious direction until darkness gathered. This morning, when he had seen the paper, he had said, and had actually believed, that the news of the murder in Basra would put an end to the trouble that had started a month ago in the Modern History class. It hadn't: the trouble, it seemed, was only beginning. And with the newspapers, and Whitburn, and Fitch, it could go on forever....

It was fully dark, now; his shadow fell ahead of him on the sidewalk, lengthening as he passed under and beyond a street-light, vanishing as he entered the stronger light of the one ahead. The windows of a cheap cafe reminded him that he was hungry, and he entered, going to a table and ordering something absently. There was a television screen over the combination bar and lunch-counter. Some kind of a comedy programme, at which an invisible studio-audience was laughing immoderately and without apparent cause. The roughly dressed customers along the counter didn't seem to see any more humor in it than he did. Then his food arrived on the table and he began to eat without really tasting it.

After a while, an alteration in the noises from the television penetrated his consciousness; a news-program had come on, and he raised his head. The screen showed a square in an Eastern city; the voice was saying:

"... Basra, where Khalid ib'n Hussein was assassinated early this morning—early afternoon, local time. This is the scene of the crime; the body of the murderer has been removed, but you can still see the stones with which he was pelted to death by the mob...."

A close-up of the square, still littered with torn-up paving-stones. A Caliphate army officer, displaying the weapon—it was an old M3, all right; Chalmers had used one of those things, himself, thirty years before, and he and his contemporaries had called it a "grease-gun." There were some recent pictures of Khalid, including one taken as he left the plane on his return from Ankara. He watched, absorbed; it was all exactly as he had "remembered" a month ago. It gratified him to see that his future "memories" were reliable in detail as well as generality.

"But the most amazing part of the story comes, not from Basra, but from Blanley College, in California," the commentator was saying, "where, it is revealed, the murder of Khalid was foretold, with uncanny accuracy, a month ago, by a history professor, Doctor Edward Chalmers...."

There was a picture of himself, in hat and overcoat, perfectly motionless, as though a brief moving glimpse were being prolonged. A glance at the background told him when and where it had been taken—a year and a half ago, at a convention at Harvard. These telecast people must save up every inch of old news-film they ever took. There were views of Blanley campus, and interviews with some of the Modern History IV boys, including Dacre and Kendrick. That was one of the things they'd been doing with that jeep-mounted sound-camera, this afternoon, then. The boys, some brashly, some embarrassedly, were substantiating the fact that he had, a month ago, described yesterday's event in detail. There was an interview with Leonard Fitch; the psychology professor was trying to explain the phenomenon of precognition in layman's terms, and making heavy going of it. And there was the mobbing of Whitburn in front of Administration Center. The college president was shouting denials of every question asked him, and as he turned and fled, the guffaws of the reporters were plainly audible.

An argument broke out along the counter.

"I don't believe it! How could anybody know all that about something before it happened?"

"Well, you heard that-there professor, what was his name. An' you heard all them boys...."

"Ah, college-boys; they'll do anything for a joke!"

"After refusing to be interviewed for telecast, the president of Blanley College finally consented to hold a press conference in his office, from which telecast cameras were barred. He denied the whole story categorically and stated that the boys in Professor Chalmers' class had concocted the whole thing as a hoax...."

"There! See what I told you!"

"... stating that Professor Chalmers is mentally unsound, and that he has been trying for years to oust him from his position on the Blanley faculty but has been unable to do so because of the provisions of the Faculty Tenure Act of 1963. Most of his remarks were in the nature of a polemic against this law, generally regarded as the college professors' bill of rights. It is to be stated here that other members of the Blanley faculty have unconditionally confirmed the fact that Doctor Chalmers did make the statements attributed to him a month ago, long before the death of Khalid ib'n Hussein...."

"Yah! How about *that*, now? How'ya gonna get around *that*?"

Beckoning the waitress, he paid his check and hurried out. Before he reached the door, he heard a voice, almost stuttering with excitement:

"Hey! Look! That's *him*!"

He began to run. He was two blocks from the cafe before he slowed to a walk again.

That night, he needed three shots of whiskey before he could get to sleep.

A delegation from the American Institute of Psionics and Parapsychology reached Blanley that morning, having taken a strato-plane from the East Coast. They had academic titles and degrees that even Lloyd Whitburn couldn't ignore. They talked with Leonard Fitch, and with the students from Modern History IV, and took statements. It wasn't until after General European History II that they caught up with Chalmers—an elderly man, with white hair and a ruddy face; a young man who looked like a heavy-weight boxer; a middle-aged man in tweeds who smoked a pipe and looked as though he ought to be more interested in grouse-shooting and flower-gardening than in clairvoyance and telepathy. The names of the first two meant nothing to Chalmers. They were important names in their own field, but it was not his field. The name of the third, who listened silently, he did not catch.

"You understand, gentlemen, that I'm having some difficulties with the college administration about this," he told them. "President Whitburn has even gone so far as to challenge my fitness to hold a position here."

"We've talked to him," the elderly man said. "It was not a very satisfactory discussion."

"President Whitburn's fitness to hold his own position could very easily be challenged," the young man added pugnaciously.

"Well, then, you see what my position is. I've consulted my attorney, Mr. Weill and he has advised me to make absolutely no statements of any sort about the matter."

"I understand," the eldest of the trio said. "But we're not the press, or anything like that. We can assure you that anything you tell us will be absolutely confidential." He looked inquiringly at the middle-aged man in tweeds, who nodded silently. "We can understand that the students in your modern history class are telling what is substantially the truth?"

"If you're thinking about that hoax statement of Whitburn's, that's a lot of idiotic drivel!" he said angrily. "I heard some of those boys on the telecast, last night; except for a few details in which they were confused, they all stated exactly what they heard me say in class a month ago."

"And we assume,"—again he glanced at the man in tweeds—"that you had no opportunity of knowing anything, at the time, about any actual plot against Khalid's life?"

The man in tweeds broke silence for the first time. "You can assume that. I don't even think this fellow Noureed knew anything about it, then."

"Well, we'd like to know, as nearly as you're able to tell us, just how you became the percipient of this knowledge of the future event of the death of Khalid ib'n Hussein," the young man began. "Was it through a dream, or a waking experience; did you visualize, or have an auditory impression, or did it simply come into your mind...."

"I'm sorry, gentlemen." He looked at his watch. "I have to be going somewhere, at once. In any case, I simply can't discuss the matter with you. I appreciate your position; I know how I'd feel if data of historical importance were being withheld from me. However, I trust that you will appreciate my position and spare me any further questioning."

That was all he allowed them to get out of him. They spent another few minutes being polite to one another; he invited them to lunch at the Faculty Club, and learned that they were lunching there as Fitch's guests. They went away trying to hide their disappointment.

The Psionics and Parapsychology people weren't the only delegation to reach Blanley that day. Enough of the trustees of the college lived in the San Francisco area to muster a quorum for a meeting the evening before; a committee, including James Dacre, the father of the boy in Modern History IV, was appointed to get the facts at first hand; they arrived about noon. They talked to some of the students, spent some time closeted with

Whitburn, and were seen crossing the campus with the Parapsychology people. They didn't talk to Chalmers or Fitch. In the afternoon, Marjorie Fenner told Chalmers that his presence at a meeting, to be held that evening in Whitburn's office, was requested. The request, she said, had come from the trustees' committee, not from Whitburn; she also told him that Fitch would be there. Chalmers promptly phoned Stanly Weill.

"I'll be there along with you," the lawyer said. "If this trustees' committee is running it, they'll realize that this is a matter in which you're entitled to legal advice. I'll stop by your place and pick you up.... You haven't been doing any talking, have you?"

He described the interview with the Psionics and Parapsychology people.

"That was all right.... Was there a man with a mustache, in a brown tweed suit, with them?"

"Yes. I didn't catch his name...."

"It's Cutler. He's an Army major; Central Intelligence. His crowd's interested in whether you had any real advance information on this. He was in to see me, just a while ago. I have the impression he'd like to see this whole thing played down, so he'll be on our side, more or less and for the time being. I'll be around to your place about eight; in the meantime, don't do any more talking than you have to. I hope we can get this straightened out, this evening. I'll have to go to Reno in a day or so to see a client there...."

The meeting in Whitburn's office had been set for eight-thirty; Weill saw to it that they arrived exactly on time. As they got out of his car at Administration Center and crossed to the steps, Chalmers had the feeling of going to a duel, accompanied by his second. The briefcase Weill was carrying may have given him the idea; it was flat and square-cornered, the size and shape of an old case of dueling pistols. He commented on it.

"Sound recorder," Weill said. "Loaded with a four-hour spool. No matter how long this thing lasts, I'll have a record of it, if I want to produce one in court."

Another party was arriving at the same time—the two Psionics and Parapsychology people and the Intelligence major, who seemed to have formed a working partnership. They all entered together, after a brief and guardedly polite exchange of greetings. There were voices raised in argument inside when they came to Whitburn's office. The college president was trying to keep Handley, Tom Smith, and Max Pottgeiter from entering his private room in the rear.

"It certainly is!" Handley was saying. "As faculty members, any controversy involving establishment of standards of fitness to teach under a tenure-contract concerns all of us, because any action taken in this case may establish a precedent which could affect the validity of our own contracts."

A big man with iron-gray hair appeared in the doorway of the private office behind Whitburn; James Dacre.

"These gentlemen have a substantial interest in this, Doctor Whitburn," he said. "If they're here as representatives of the college faculty, they have every right to be present."

Whitburn stood aside. Handley, Smith and Pottgeiter went through the door; the others followed. The other three members of the trustees' committee were already in the room. A few minutes later, Leonard Fitch arrived, also carrying a briefcase.

"Well, everybody seems to be here," Whitburn said, starting toward his chair behind the desk. "We might as well get this started."

"Yes. If you'll excuse me, Doctor." Dacre stepped in front of him and sat down at the desk. "I've been selected as chairman of this committee; I believe I'm presiding here. Start the recorder, somebody."

One of the other trustees went to the sound recorder beside the desk—a larger but probably not more efficient instrument than the one Weill had concealed in his briefcase—and flipped a switch. Then he and his companions dragged up chairs to flank Dacre's, and the rest seated themselves around the room. Old Pottgeiter took a seat next to Chalmers. Weill opened the case on his lap, reached inside, and closed it again.

"What are they trying to do, Ed?" Pottgeiter asked, in a loud whisper. "Throw you off the faculty? They can't do that, can they?"

"I don't know, Max. We'll see...."

"This isn't any formal hearing, and nobody's on trial here," Dacre was saying. "Any action will have to be taken by the board of trustees as a whole, at a regularly scheduled meeting. All we're trying to do is find out just what's happened here, and who, if anybody, is responsible...."

"Well, there's the man who's responsible!" Whitburn cried, pointing at Chalmers. "This whole thing grew out of his behavior in class a month ago, and I'll remind you that at the time I demanded his resignation!"

"I thought it was Doctor Fitch, here, who gave the story to the newspapers," one of the trustees, a man with red hair and a thin, eyeglassed face, objected.

"Doctor Fitch acted as any scientist should, in making public what he believed to be an important scientific discovery," the elder of the two Parapsychology men said. "He believed, and so do we, that he had discovered a significant instance of precognition—a case of real prior knowledge of a future event. He made a careful and systematic record of Professor Chalmers' statements, at least two weeks before the occurrence of the event to which they referred. It is entirely due to him that we know exactly what Professor Chalmers said and when he said it."

"Yes," his younger colleague added, "and in all my experience I've never heard anything more preposterous than this man Whitburn's attempt, yesterday, to deny the fact."

"Well, we're convinced that Doctor Chalmers did in fact say what he's alleged to have said, last month," Dacre began.

"Jim, I think we ought to get that established, for the record," another of the trustees put in. "Doctor Chalmers, is it true that you spoke, in the past tense, about the death of Khalid ib'n Hussein in one of your classes on the sixteenth of last month?"

Chalmers rose. "Yes, it is. And the next day, I was called into this room by Doctor Whitburn, who demanded my resignation from the faculty of this college because of it. Now, what I'd like to know is, why did Doctor Whitburn, in this same room, deny, yesterday, that I'd said anything of the sort, and accuse my students of concocting the story after the event as a hoax."

"One of them being my son," Dacre added. "I'd like to hear an answer to that, myself."

"So would I," Stanly Weill chimed in. "You know, my client has a good case against Doctor Whitburn for libel."

Chalmers looked around the room. Of the thirteen men around him, only Whitburn was an enemy. Some of the others were on his side, for one reason or another, but none of them were friends. Weill was his lawyer, obeying an obligation to a client which, at bottom, was an obligation to his own conscience. Handley was afraid of the possibility that a precedent might be established which would impair his own tenure-contract. Fitch, and the two men from the Institute of Psionics and Parapsychology were interested in him as a source of study-material. Dacre resented a slur upon his son; he and the others were interested in Blanley College as an institution, almost an abstraction. And the major in mufti was probably worrying about the consequences to military security of having a prophet at large. Then a hand gripped his shoulder, and a voice whispered in his ear:

"That's good, Ed; don't let them scare you!"

Old Max Pottgeiter, at least, was a friend.

"Doctor Whitburn, I'm asking you, and I expect an answer, why did you make such statements to the press, when you knew perfectly well that they were false?" Dacre demanded sharply.

"I knew nothing of the kind!" Whitburn blustered, showing, under the bluster, fear. "Yes, I demanded this man's resignation on the morning of October Seventeenth, the day after this incident occurred. It had come to my attention on several occasions that he was making wild and unreasonable assertions in class, and subjecting himself, and with himself the whole faculty of this college, to student ridicule. Why, there was actually an editorial about it written by the student editor of the campus paper, the *Black and Green*. I managed to prevent its publication...." He went on at some length about that. "If I might be permitted access to the drawers of my own desk," he added with elephantine sarcasm, "I could show you the editorial in question."

"You needn't bother; I have a carbon copy," Dacre told him. "We've all read it. If you did, at the time you suppressed it, you should have known what Doctor Chalmers said in class."

"I knew he'd talked a lot of poppycock about a man who was still living having been shot to death," Whitburn retorted. "And if something of the sort actually happened, what of it? Somebody's always taking a shot at one or another of these foreign dictators, and they can't miss all the time."

"You claim this was pure coincidence?" Fitch demanded. "A ten-point coincidence: Event of assassination, year of the event, place, circumstances, name of assassin, nationality of assassin, manner of killing, exact type of weapon used, guards killed and wounded along with Khalid, and fate of the assassin. If that's a simple and plausible coincidence, so's dealing ten royal flushes in succession in a poker game. Tom, you figured that out; what did you say the odds against it were?"

"Was all that actually stated by Doctor Chalmers a month ago?" one of the trustees asked, incredulously.

"It absolutely was. Look here, Mr. Dacre, gentlemen." Fitch came forward, unzipping his briefcase and pulling out papers. "Here are the signed statements of each of Doctor Chalmers' twenty-three Modern History Four students, all made and dated before the assassination. You can refer to them as you please; they're in alphabetical order. And here." He unfolded a sheet of graph paper a yard long and almost as wide. "Here's a tabulated summary of the boys' statements. All agreed on the first point, the fact of the assassination. All agreed that the time was sometime this year. Twenty out of twenty-three agreed on Basra as the place. Why, seven of them even remembered the name of the assassin. That in itself is remarkable; Doctor Chalmers has an extremely intelligent and attentive class."

"They're attentive because they know he's always likely to do something crazy and make a circus out of himself," Whitburn interjected.

"And this isn't the only instance of Doctor Chalmers' precognitive ability," Fitch continued. "There have been a number of other cases...."

Chalmers jumped to his feet; Stanly Weill rose beside him, shoved the cased sound-recorder into his hands, and pushed him back into his seat.

"Gentlemen," the lawyer began, quietly but firmly and clearly. "This is all getting pretty badly out of hand. After all, this isn't an investigation of the actuality of precognition as a psychic phenomenon. What I'd like to hear, and what I haven't heard yet, is Doctor Whitburn's explanation of his contradictory statements that he knew about my client's alleged remarks on the evening after they were supposed to have been made and that, at the same time, the whole thing was a hoax concocted by his students."

"Are you implying that I'm a liar?" Whitburn bristled.

"I'm pointing out that you made a pair of contradictory statements, and I'm asking how you could do that knowingly and honestly," Weill retorted.

"What I meant," Whitburn began, with exaggerated slowness, as though speaking to an idiot, "was that yesterday, when those infernal reporters were badgering me, I really thought that some of Professor Chalmers' students had gotten together and given the *Valley Times* an exaggerated story about his insane maunderings a month ago. I hadn't imagined that a member of the faculty had been so lacking in loyalty to the college...."

"You couldn't imagine anybody with any more intellectual integrity than you have!" Fitch fairly yelled at him.

"You're as crazy as Chalmers!" Whitburn yelled back. He turned to the trustees. "You see the position I'm in, here, with this infernal Higher Education Faculty Tenure Act? I have a madman on my faculty, and can I get rid of him? No! I demand his resignation, and he laughs at me and goes running for his lawyer! And he is a madman! Nobody but a madman would talk the way he does. You think this Khalid ib'n Hussein business is the only time he's done anything like this? Why, I have a list of a dozen occasions when he's done something just as bad, only he didn't have a lucky coincidence to back him up. Trying to get books that don't exist out of the library, and then insisting that they're standard textbooks. Talking about the revolt of the colonies on Mars and Venus. Talking about something he calls the Terran Federation, some kind of a world empire. Or something he calls Operation Triple Cross, that saved the country during some fantastic war he imagined...."

"*What did you say?*"

The question cracked out like a string of pistol shots. Everybody turned. The quiet man in the brown tweed suit had spoken; now he looked as though he were very much regretting it.

"Is there such a thing as Operation Triple Cross?" Fitch was asking.

"No, no. I never heard anything about that; that wasn't what I meant. It was this Terran Federation thing," the major said, a trifle too quickly and too smoothly. He turned to Chalmers. "You never did any work for PSPB; did you ever talk to anybody who did?" he asked.

"I don't even know what the letters mean," Chalmers replied.

"Politico-Strategic Planning Board. It's all pretty hush-hush, but this term Terran Federation is a tentative name for a proposed organization to take the place of the U. N. if that organization breaks up. It's nothing particularly important, and it only exists on paper."

It won't exist only on paper very long, Chalmers thought. He was wondering what Operation Triple Cross was; he had some notes on it, but he had forgotten what they were.

"Maybe he did pick that up from somebody who'd talked indiscreetly," Whitburn conceded. "But the rest of this tommyrot! Why, he was talking about how the city of Reno had been destroyed by an explosion and fire, literally wiped off the map. There's an example for you!"

He'd forgotten about that, too. It had been a relatively minor incident in the secret struggle of the Subwar; now he remembered having made a note about it. He was sure that it followed closely after the assassination of Khalid ib'n Hussein. He turned quickly to Weill.

"Didn't you say you had to go to Reno in a day or so?" he asked.

Weill hushed him urgently, pointing with his free hand to the recorder. The exchange prevented him from noticing that Max Pottgeiter had risen, until the old man was speaking.

"Are you trying to tell these people that Professor Chalmers is crazy?" he was demanding. "Why, he has one of the best minds on the campus. I was talking to him only yesterday, in the back room at the Library. You know," he went on apologetically, "my subject is Medieval History; I don't pay much attention to what's going on in the contemporary world, and I didn't understand, really, what all this excitement was about. But he explained the whole thing to me, and did it in terms that I could grasp, drawing some excellent parallels with the Byzantine Empire and the Crusades. All about the revolt at Damascus, and the sack of Beirut, and the war between Jordan and Saudi Arabia, and how the Turkish army intervened, and the invasion of Pakistan...."

"When did all this happen?" one of the trustees demanded.

Pottgeiter started to explain; Chalmers realized, sickly, how much of his future history he had poured into the trusting ear of the old medievalist, the day before.

"Good Lord, man; don't you read the papers at all?" another of the trustees asked.

"No! And I don't read inside-dope magazines, or science fiction. I read carefully substantiated facts. And I know when I'm talking to a sane and reasonable man. It isn't a common experience, around here."

Dacre passed a hand over his face. "Doctor Whitburn," he said, "I must admit that I came to this meeting strongly prejudiced against you, and I'll further admit that your own behavior here has done very little to dispel that prejudice. But I'm beginning to get some idea of what you have to contend with, here at Blanley, and I find that I must make a lot of allowances. I had no idea.... Simply no idea at all."

"Look, you're getting a completely distorted picture of this, Mr. Dacre," Fitch broke in. "It's precisely as I believed; Doctor Chalmers is an unusually gifted precognitive percipient. You've seen, gentlemen, how his complicated chain of precognitions about the death of Khalid has been proven veridical; I'd stake my life that every one of these precognitions will be similarly verified. And I'll stake my professional reputation that the man is perfectly sane. Of course, abnormal psychology and psychopathology aren't my subjects, but...."

"They're not my subjects, either," Whitburn retorted, "but I know a lunatic by his ravings."

"Doctor Fitch is taking an entirely proper attitude," Pottgeiter said, "in pointing out that abnormal psychology is a specialized branch, outside his own field. I wouldn't dream, myself, of trying to offer a decisive opinion on some point of Roman, or Babylonian, history. Well, if the question of Doctor Chalmers' sanity is at issue here, let's consult somebody who specializes in insanity. I don't believe that anybody here is qualified even to express an opinion on that subject, Doctor Whitburn least of all."

Whitburn turned on him angrily. "Oh, shut up, you doddering old fool!" he shouted. "Look; there's another of them!" he told the trustees. "Another deadhead on the faculty that this Tenure Law keeps me from getting rid of. He's as bad as Chalmers, himself. You just heard that string of nonsense he was spouting. Why, his courses have been noted among the students for years as snap courses in which nobody ever has to do any work...."

Chalmers was on his feet again, thoroughly angry. Abuse of himself he could take; talking that way about gentle, learned, old Pottgeiter was something else.

"I think Doctor Pottgeiter's said the most reasonable thing I've heard since I came in here," he declared. "If my sanity is to be questioned, I insist that it be questioned by somebody qualified to do so."

Had the sane restrained the insane, or was it the other way around?

Weill set his recorder on the floor and jumped up beside him, trying to haul him back into his seat.

"For God's sake, man! Sit down and shut up!" he hissed.

Chalmers shook off his hand. "No, I won't shut up! This is the only way to settle this, once and for all. And when my sanity's been vindicated, I'm going to sue this fellow...."

Whitburn started to make some retort, then stopped short. After a moment, he smiled nastily.

"Do I understand, Doctor Chalmers, that you would be willing to submit to psychiatric examination?" he asked.

"Don't agree; you're putting your foot in a trap!" Weill told him urgently.

"Of course, I agree, as long as the examination is conducted by a properly qualified psychiatrist."

"How about Doctor Hauserman at Northern State Mental Hospital?" Whitburn asked quickly. "Would you agree to an examination by him?"

"Excellent!" Fitch exclaimed. "One of the best men in the field. I'd accept his opinion unreservedly."

Weill started to object again; Chalmers cut him off. "Doctor Hauserman will be quite satisfactory to me. The only question is, would he be available?"

"I think he would," Dacre said, glancing at his watch. "I wonder if he could be reached now." He got to his feet. "Telephone in your outer office, Doctor Whitburn? Fine. If you gentlemen will excuse me...."

It was a good fifteen minutes before he returned, smiling.

"Well, gentlemen, it's all arranged," he said. "Doctor Hauserman is quite willing to examine Doctor Chalmers—with the latter's consent, of course."

"He'll have it. In writing, if he wishes."

"Yes, I assured him on that point. He'll be here about noon tomorrow—it's a hundred and fifty miles from the hospital, but the doctor flies his own plane—and the examination can start at two in the afternoon. He seems familiar with the facilities of the psychology department, here; I assured him that they were at his disposal. Will that be satisfactory to you, Doctor Chalmers?"

"I have a class at that time, but one of the instructors can take it over—if holding classes will be possible around here tomorrow," he said. "Now, if you gentlemen will pardon me, I think I'll go home and get some sleep."

Weill came up to the apartment with him. He mixed a couple of drinks and they went into the living room with them.

"Just in case you don't know what you've gotten yourself into," Weill said, "this Hauserman isn't any ordinary couch-pilot; he's the state psychiatrist. If he gets the idea you aren't sane, he can commit you to a hospital, and I'll bet that's exactly what Whitburn had in mind when he suggested him. And I don't trust this man Dacre. I thought he was on our side, at the start, but that was before your friends got into the act." He frowned into his drink. "And I don't like the way that Intelligence major was acting, toward the last. If he thinks you know something you are not supposed to, a mental hospital may be his idea of a good place to put you away."

"You don't think this man Hauserman would allow himself to be influenced ...? No. You just don't think I'm sane. Do you?"

"I know what Hauserman'll think. He'll think this future history business is a classical case of systematized schizoid delusion. I wish I'd never gotten into this case. I wish I'd never even heard of you! And another thing; in case you get past Hauserman all right, you can forget about that damage-suit bluff of mine. You would not stand a chance with it in court."

"In spite of what happened to Khalid?"

"After tomorrow, I won't stay in the same room with anybody who even mentions that name to me. Well, win or lose, it'll be over tomorrow and then I can leave here."

"Did you tell me you were going to Reno?" Chalmers asked. "Don't do it. You remember Whitburn mentioning how I spoke about an explosion there? It happened just a couple of days after the murder of Khalid. There was—will be—a trainload of high explosives in the railroad yard; it'll be the biggest non-nuclear explosion since the *Mont Blanc* blew up in Halifax harbor in World War One...."

Weill threw his drink into the fire; he must have avoided throwing the glass in with it by a last-second exercise of self-control.

"Well," he said, after a brief struggle to master himself. "One thing about the legal profession; you do hear the damnedest things!... Good night, Professor. And try—please try, for the sake of your poor harried lawyer—to keep your mouth shut about things like that, at least till after you get through with Hauserman. And when you're talking to him, don't, don't, for heaven's sake, *don't*, volunteer anything!"

The room was a pleasant, warmly-colored, place. There was a desk, much like the ones in the classrooms, and six or seven wicker armchairs. A lot of apparatus had been pushed back along the walls; the dust-covers were gay cretonne. There was a couch, with more apparatus, similarly covered, beside it. Hauserman was seated at the desk when Chalmers entered.

He rose, and they shook hands. A man of about his own age, smooth-faced, partially bald. Chalmers tried to guess something of the man's nature from his face, but could read nothing. A face well trained to keep its owner's secrets.

"Something to smoke, Professor," he began, offering his cigarette case.

"My pipe, if you don't mind." He got it out and filled it.

"Any of those chairs," Hauserman said, gesturing toward them.

They were all arranged to face the desk. He sat down, lighting his pipe. Hauserman nodded approvingly; he was behaving calmly, and didn't need being put at ease. They talked at random—at least, Hauserman tried to make it seem so—for some time about his work, his book about the French Revolution, current events. He picked his way carefully through the conversation, alert for traps which the psychiatrist might be laying for him. Finally, Hauserman said:

"Would you mind telling me just why you felt it advisable to request a psychiatric examination, Professor?"

"I didn't request it. But when the suggestion was made, by one of my friends, in reply to some aspersions of my sanity, I agreed to it."

"Good distinction. And why was your sanity questioned? I won't deny that I had heard of this affair, here, before Mr. Dacre called me, last evening, but I'd like to hear your version of it."

He went into that, from the original incident in Modern History IV, choosing every word carefully, trying to concentrate on making a good impression upon Hauserman, and at the same time finding that more "memories" of the future were beginning to seep past the barrier of his consciousness. He tried to dam them back; when he could not, he spoke with greater and greater care lest they leak into his speech.

"I can't recall the exact manner in which I blundered into it. The fact that I did make such a blunder was because I was talking extemporaneously and had wandered ahead of my text. I was trying to show the results of the collapse of the Ottoman Empire after the First World War, and the partition of the Middle East into a loose collection of Arab states, and the passing of British and other European spheres of influence following the Second. You know, when you consider it, the Islamic Caliphate was inevitable; the surprising thing is that it was created by a man like Khalid...."

He was talking to gain time, and he suspected that Hauserman knew it. The "memories" were coming into his mind more and more strongly; it was impossible to suppress them. The period of anarchy following Khalid's death would be much briefer, and much more violent, than he had previously thought. Tallal ib'n Khalid would be flying from England even now; perhaps he had already left the plane to take refuge among the black tents of his father's Bedouins. The revolt at Damascus would break out before the end of the month; before the end of the year, the whole of Syria and Lebanon would be in bloody chaos, and the Turkish army would be on the march.

"Yes. And you allowed yourself to be carried a little beyond the present moment, into the future, without realizing it? Is that it?"

"Something like that," he replied, wide awake to the trap Hauserman had set, and fearful that it might be a blind, to disguise the real trap. "History follows certain patterns. I'm not a Toynbean, by any manner of means, but any historian can see that certain forces generally tend to produce similar effects. For instance, space travel is now a fact; our government has at present a military base on Luna. Within our lifetimes—certainly within the lifetimes of my students—there will be explorations and attempts at colonization on Mars and Venus. You believe that, Doctor?"

"Oh, unreservedly. I'm not supposed to talk about it, but I did some work on the Philadelphia Project, myself. I'd say that every major problem of interplanetary flight had been solved before the first robot rocket was landed on Luna."

"Yes. And when Mars and Venus are colonized, there will be the same historic situations, at least in general shape, as arose when the European powers were colonizing the New World, or, for that matter, when the Greek city-states were throwing out colonies across the Aegean. That's the sort of thing we call projecting the past into the future through the present."

Hauserman nodded. "But how about the details? Things like the assassination of a specific personage. How can you extrapolate to a thing like that?"

"Well...." More "memories" were coming to the surface; he tried to crowd them back. "I do my projecting in what you might call fictionalized form; try to fill in the details from imagination. In the case of Khalid, I was trying to imagine what would happen if his influence were suddenly removed from Near Eastern and Middle Eastern, affairs. I suppose I constructed an imaginary scene of his assassination...."

He went on at length. Mohammed and Noureed were common enough names. The Middle East was full of old U. S. weapons. Stoning was the traditional method of execution; it diffused responsibility so that no individual could be singled out for blood-feud vengeance.

"You have no idea how disturbed I was when the whole thing happened, exactly as I had described it," he continued. "And worst of all, to me, was this Intelligence officer showing up; I thought I was really in for it!"

"Then you've never really believed that you had real knowledge of the future?"

"I'm beginning to, since I've been talking to these Psionics and Parapsychology people," he laughed. It sounded, he hoped, like a natural and unaffected laugh. "They seem to be convinced that I have."

There would be an Eastern-inspired uprising in Azerbaijan by the middle of the next year; before autumn, the Indian Communists would make their fatal attempt to seize East Pakistan. The Thirty Days' War would be the immediate result. By that time, the Lunar Base would be completed and ready; the enemy missiles would be aimed primarily at the rocketports from which it was supplied. Delivered without warning, it should have succeeded—except that every rocketport had its secret duplicate and triplicate. That was Operation Triple Cross; no wonder Major Cutler had been so startled at the words, last evening. The enemy would be utterly overwhelmed under the rain of missiles from across space, but until the moon-rockets began to fall, the United States would suffer grievously.

"Honestly, though, I feel sorry for my friend Fitch," he added. "He's going to be frightfully let down when some more of my alleged prophecies misfire on him. But I really haven't been deliberately deceiving him."

And Blanley College was at the center of one of the areas which would receive the worst of the thermonuclear hell to come. And it would be a little under a year....

"And that's all there is to it!" Hauserman exclaimed, annoyance in his voice. "I'm amazed that this man Whitburn allowed a thing like this to assume the proportions it did. I must say that I seem to have gotten the story about this business in a very garbled form indeed." He laughed shortly. "I came here convinced that you were mentally unbalanced. I hope you won't take that the wrong way, Professor," he hastened to add. "In my profession, anything can be expected. A good psychiatrist can never afford to forget how sharp and fine is the knife-edge."

"The knife-edge!" The words startled him. He had been thinking, at that moment, of the knife-edge, slicing moment after moment relentlessly away from the future, into the past, at each slice coming closer and closer to the moment when the missiles of the Eastern Axis would fall. "I didn't know they still resorted to surgery, in mental cases," he added, trying to cover his break.

"Oh, no; all that sort of thing is as irrevocably discarded as the whips and shackles of Bedlam. I meant another kind of knife-edge; the thin, almost invisible, line which separates sanity from non-sanity. From madness, to use a deplorable lay expression."

Hauserman lit another cigarette. "Most minds are a lot closer to it than their owners suspect, too. In fact, Professor, I was so convinced that yours had passed over it that I brought with me a commitment form, made out all but my signature, for you." He took it from his pocket and laid it on the desk. "The modern equivalent of the *lettre-de-cachet*, I suppose the author of a book on the French Revolution would call it. I was all ready to certify you as mentally unsound, and commit you to Northern State Mental Hospital."

Chalmers sat erect in his chair. He knew where that was; on the other side of the mountains, in the one part of the state completely untouched by the H-bombs of the Thirty Days' War. Why, the town outside which the hospital stood had been a military headquarters during the period immediately after the bombings, and the center from which all the rescue work in the state had been directed.

"And you thought you could commit me to Northern State!" he demanded, laughing scornfully, and this time he didn't try to make the laugh sound natural and unaffected. "You—confine *me*, anywhere? Confine a poor old history professor's body, yes, but that isn't me. I'm universal; I exist in all space-time. When this old body I'm wearing now was writing that book on the French Revolution, I was in Paris, watching it happen, from the fall of the Bastile to the Ninth Thermidor. I was in Basra, and saw that crazed tool of the Axis shoot down Khalid ib'n Hussein—and the professor talked about it a month before it happened. I have seen empires rise and stretch from star to star across the Galaxy, and crumble and fall. I have seen...."

Doctor Hauserman had gotten his pen out of his pocket and was signing the commitment form with one hand; with the other, he pressed a button on the desk. A door at the rear opened, and a large young man in a white jacket entered.

"You'll have to go away for a while, Professor," Hauserman was telling him, much later, after he had allowed himself to become calm again. "For how long, I don't know. Maybe a year or so."

"You mean to Northern State Mental?"

"Well.... Yes, Professor. You've had a bad crack-up. I don't suppose you realize how bad. You've been working too hard; harder than your nervous system could stand. It's been too much for you."

"You mean, I'm nuts?"

"Please, Professor. I deplore that sort of terminology. You've had a severe psychological breakdown...."

"Will I be able to have books, and papers, and work a little? I couldn't bear the prospect of complete idleness."

"That would be all right, if you didn't work too hard."

"And could I say good-bye to some of my Friends?

Hauserman nodded and asked, "Who?"

"Well, Professor Pottgeiter...."

"He's outside now. He was inquiring about you."

"And Stanly Weill, my attorney. Not business; just to say good-bye."

"Oh, I'm sorry, Professor. He's not in town, now. He left almost immediately after.... After...."

"After he found out I was crazy for sure? Where'd he go?"

"To Reno; he took the plane at five o'clock."

Weill wouldn't have believed, anyhow; no use trying to blame himself for that. But he was as sure that he would never see Stanly Weill alive again as he was that the next morning the sun would rise. He nodded impassively.

"Sorry he couldn't stay. Can I see Max Pottgeiter alone?"

"Yes, of course, Professor."

Old Pottgeiter came in, his face anguished. "Ed! It isn't true," he stammered. "I won't believe that it's true."

"What, Max?"

"That you're crazy. Nobody can make me believe that."

He put his hand on the old man's shoulder. "Confidentially, Max, neither do I. But don't tell anybody I'm not. It's a secret."

Pottgeiter looked troubled. For a moment, he seemed to be wondering if he mightn't be wrong and Hauserman and Whitburn and the others right.

"Max, do you believe in me?" he asked. "Do you believe that I knew about Khalid's assassination a month before it happened?"

"It's a horribly hard thing to believe," Pottgeiter admitted. "But, dammit, Ed, you did! I know, medieval history is full of stories about prophecies being fulfilled. I always thought those stories were just legends that grew up after the event. And, of course, he's about a century late for me, but there was Nostradamus. Maybe those old prophecies weren't just *ex post facto* legends, after all. Yes. After Khalid, I'll believe that."

"All right. I'm saying, now, that in a few days there'll be a bad explosion at Reno, Nevada. Watch the papers and the telecast for it. If it happens, that ought to prove it. And you remember what I told you about the Turks annexing Syria and Lebanon?" The old man nodded. "When that happens, get away from Blanley. Come up to the town where Northern State Mental Hospital is, and get yourself a place to live, and stay there. And try to bring Marjorie Fenner along with you. Will you do that, Max?"

"If you say so." His eyes widened. "Something bad's going to happen here?"

"Yes, Max. Something very bad. You promise me you will?"

"Of course, Ed. You know, you're the only friend I have around here. You and Marjorie. I'll come, and bring her along."

"Here's the key to my apartment." He got it from his pocket and gave it to Pottgeiter, with instructions. "Everything in the filing cabinet on the left of my desk. And don't let anybody else see any of it. Keep it safe for me."

The large young man in the white coat entered.

THE END